Heitor Luiz Borali

Effects of Global Warming on Coral Reefs

Heitor Luiz Borali

Effects of Global Warming on Coral Reefs

A Warning about Environmental Degradation

ScienciaScripts

Imprint

Any brand names and product names mentioned in this book are subject to trademark, brand or patent protection and are trademarks or registered trademarks of their respective holders. The use of brand names, product names, common names, trade names, product descriptions etc. even without a particular marking in this work is in no way to be construed to mean that such names may be regarded as unrestricted in respect of trademark and brand protection legislation and could thus be used by anyone.

Cover image: www.ingimage.com

This book is a translation from the original published under ISBN 978-613-9-75116-7.

Publisher:
Sciencia Scripts
is a trademark of
Dodo Books Indian Ocean Ltd. and OmniScriptum S.R.L publishing group

120 High Road, East Finchley, London, N2 9ED, United Kingdom
Str. Armeneasca 28/1, office 1, Chisinau MD-2012, Republic of Moldova, Europe
Printed at: see last page
ISBN: 978-620-6-42407-9

DEDICATORY

To my parents,

Luzia Ribeiro Borali and Gerson Borali

For your support throughout my life, for the trust you have always placed in me and, above all, for your endless love, attention and dedication. You are very special in my life...

To my brothers,

Glauciléia Borali and Wendel Borali

For your love, friendship, understanding and so much help. I love you...

To my nephews,

Brendon Breda Borali and Nathália Borali Boldori

Precious things in all our lives. I love you little ones...

To my great love,

Luiz Fernando Galhardo

For love, help in difficult times and eternal understanding.

ACKNOWLEDGEMENTS

To Iguaçú University

For the opportunity to train and carry out this work.

To my teachers,

Who have taught me so much and got me this far...

To my family and everyone who, directly or indirectly, contributed to the realisation of this work, to my education and to my life.

THANK YOU FOR EVERYTHING!!!

"When the last tree has fallen, when the last river has dried up, when the last fish has been caught, you'll realise that money can't be eaten."

Greenpeace

TABLE OF CONTENTS:

INTRODUCTION

One of the biggest environmental problems that has attracted the attention not only of the entire scientific community, but of everyone, is global warming, which is causing consequences all over the globe, all due to the increased concentration of GHGs (greenhouse gases), the main gas being carbon dioxide (CO_2), which is emitted in excess by cars, factories, fuel burning and even natural sources in the environment. As a result, the accumulation of GHGs in the atmosphere causes the intensification of a natural phenomenon called the greenhouse effect.

The greenhouse effect is perhaps the most global of the many threats to the planet's biodiversity. Unlike others caused by human development, it has the potential to impact all ecosystems, including those far from human populations and still classified as unexplored. The possibility of extensive impacts implies a direct threat to the planet's biodiversity. This refers to the physical process by which the presence of atmospheric gases causes the planet to maintain a higher equilibrium temperature than it would if these gases were absent, allowing light to pass through and trapping heat.

Despite the controversies surrounding the causes of global warming, the information gathered throughout this century and especially in the last decade has shown that human activities have been able to introduce global changes that could have consequences serious enough to affect natural ecosystems. Thus, global warming is generating a series of serious consequences, especially for the oceans, such as rising sea levels, rising temperatures, altered sea currents, among other factors that are already contributing to affecting the entire ocean and its marine species, leaving marine fauna and flora in danger.

However, the biggest concern is the consequences of global warming on coral reefs, which are large ecosystems known as the nurseries of the seas due to their great species richness and are considered to be the most diverse marine habitat. There are approximately 350 known species of coral, which are home to around 30 per cent of existing marine species and 65 per cent of fish. As well as being a great tourist attraction, they are very important and protect the sea coast. One of the most obvious effects of global warming on reefs is coral bleaching, which has actually been used as a parameter to monitor climate change on the planet. In addition to bleaching, other warming factors such as erosion and sedimentation on the coasts and, above all, the high concentrations of CO_2 that the oceans absorb have been causing a series of problems for corals, These corals are very sensitive to stress, which can worsen the situation and even lead to their death, which would affect all the animals that depend on corals and the entire marine chain, thus influencing the entire reef.

It is estimated that the oceans are not infinite reservoirs of CO_2 and that this rate of absorption can decline over time, which, depending on the concentration, can cause further harm to corals and marine biomes.

However, concern for the preservation of coral reefs has increased greatly, as this ecosystem has suffered so much, and discovering better forms and techniques of environmental conservation, ways to reduce the problems mentioned and, above all, raising awareness of this major problem have been the main goals for the preservation of coral reefs.

Finally, this work focuses on showing the whole problem of one of the worst phenomena today, global warming, and its various consequences on the oceans and, above all, on coral reefs, the most important ecosystem in the marine environment, bringing certain benefits and harms, trying to show the importance of preserving this environment, aiming to discuss the tolerance of these organisms and proposing solutions to the issue developed.

Despite the many dangers that threaten the balance of the biosphere, there is still time for humanity to realise the need for a rational ordering of the Earth and take responsibility for implementing a strategy for a stable society. The integrity of the planet depends on just one condition: respect for the ecological laws to which man, as an element of the biosphere, is subject.

CHAPTER 1
GLOBAL CLIMATE CHANGE

1.1 CLIMATE CHANGE

The influence of climate on physical and topographical aspects plays a direct role in determining which organism has the genetic basis to be able to tolerate the climatic variations of that location and therefore be able to live there (Wilson & Frances, 1997).

Thus, organisms previously housed in a particular location, under a certain pre-established climatic action, would not be able to adapt as quickly to a new location under the action of climate change (Simon & DeFries, 1992).

Climate change occurs due to internal and external factors. Internal factors are those associated with the complexity of climate systems being chaotic non-linear systems (Gralla, 1998). External factors can be natural or anthropogenic. External factors of an anthropogenic nature have been of great relevance to the climatic events experienced in the last century (CNI - Confederação Nacional das Indústrias).

In the last millennium two important periods of temperature variation occurred: a warm period known as the Medieval Warm Period and a cold period known as the Little Ice Age. The temperature variation of these periods is similar in magnitude to the current warming and is believed to have been caused by both internal and external factors (Demillo, 1998).

The Little Ice Age is attributed to a reduction in solar activity and some scientists agree that the terrestrial warming observed since 1860 is a natural reversal of the Little Ice Age (Demillo 1998). The geological record shows that the last act of the glacial drama began when the most recent glaciation began to wane approximately 18,000 years ago. As has always been the pattern, the cold period lasted approximately 100,000 years; the current pleasant climate is a brief warm season of a characteristically frigid cycle (Simon & DeFries, 1992).

This recent shift from a glacial to a warm phase is of particular interest to the scientific community, which is struggling to understand the complexities of the modern climate, because the amount of carbon dioxide that has accumulated in the atmosphere from the beginning of the melting period to the present is more or less equal to the amount of greenhouse gases that are projected to accumulate in the atmosphere from now until approximately the middle of the next century (Simon & DeFries, 1992).

The climate has also changed over the last few centuries and decades. Although these changes have not been as drastic as those that occurred in previous periods, nor as great as those that are expected during the 21st century, it is known precisely when they began and when they ended. From this data, scientists know that the climate can change suddenly and that the changes can be large enough to cause regional impacts (Simon & DeFries, 1992).

The natural variability of the climate means that it is difficult to recognise the early stages of man-made climate change. Man has produced changes in geological forces on the planet he inhabits that would leave the Earth on the verge of climate change at a speed unprecedented in history (Wilson & Frances, 1997).

Climate change is nothing new on Earth, which has been undergoing alterations as a result of astronomical factors such as variations in the intensity of the energy emitted by the Sun, oscillations in the planet's orbital parameters around the Sun and variations in the concentrations of the chemical composition of the atmosphere (Simon & DeFries,1992).

In the last century, the planet's average temperature has risen by around 0.7°C. By simulating climate models, in which different concentrations of greenhouse gases are taken into account, it is shown that human activities can also alter current climatic conditions.

When scientists say that, on average, the global temperature could rise by a few degrees centigrade, they are talking about a considerable amount of heat. The current average temperature of the globe is approximately 14°C. An increase of 3°C would create conditions that some organisms haven't had to live with in the last 100,000 years. If the temperature rises by 4°C, the Earth could be warmer than at any time since the Eocene period 40 million years ago (Simon & DeFries, 1992).

What appears to be a small increase in temperature can have quite drastic effects. The increase in major climatic phenomena will most likely occur not in the total number of violent climatic events, but in the ability of human beings to detect and record these events (Demillo, 1998).

There is significant evidence that violent weather activity has not even increased (Demillo, 1998). This human tendency to view recent weather reports as long-term trends contrasts with the ability to predict, detect and monitor these events more frequently, so the perception is that these events are increasing (Simon & DeFries, 1992).

Data collection in less technologically favoured countries contrasts with the advanced techniques of the United States and other technologically advanced countries. Global monitoring systems, geostationary satellites, do indeed collect information from other areas of the planet, but ground stations, radars and other local detection techniques are unfortunately inadequate outside the first world. Not only is this a disservice to those living in these countries, but it leaves gaps in the global climate database (Demillo, 1998).

The planet and its delicate climate system have mechanisms for regulating and controlling the production and removal of atmospheric gases (Mazza & Roth, 1999). Humans, however, have the capacity to produce atmospheric pollutants or inhibit mechanisms that control atmospheric regulation on an enormous scale in a relatively short space of time (Becker *et al.*, 2001).

1.2 EVIDENCE OF GLOBAL WARMING

Despite the controversy surrounding the causes of global warming, the effects it has had on the planet are indisputable. In the last 50 years, an increase in temperature of around 1.5 °C has been detected in Antarctica, which has caused glaciers to melt and plants to appear that didn't exist there in the past. According to forecasts, if global warming is not controlled, there will be further rises in sea levels, which could lead to a greater incidence of tsunamis, hurricanes and tornadoes (IPCC, 2001b).

Due to ocean-atmosphere interaction, there will be negative impacts on coral reefs, caused by high levels of carbon dioxide in the atmosphere. Mangroves, maritime vegetation, other coastal ecosystems and associated biodiversity will be affected by rising temperatures and accelerated sea level rise (IPCC, 2001b).

Due to the increase in temperature, the atmosphere will become more humid in response, increasing the level of precipitation in tropical regions. This also increases the number of insects and the diseases they cause and waterborne diseases (IPCC, 2001b).

Considering the scenario of rising temperatures, it can be assumed that in climatic regions bordering those where agricultural plants can be grown, the change that occurs will be unfavourable to plant development. The greater the anomaly, the less adapted the region will become, up to the maximum limit of biological tolerance to heat. On the other hand, other crops that are more resistant to high temperatures will probably benefit, up to their own limit of tolerance to heat stress (IPCC, 2001b).

In countries like Canada, Iceland and Sweden, for example, the effects of rising temperatures could be beneficial. Soils that are not suitable for agricultural cultivation could become so (IPCC, 2001b).

In the case of low temperatures, regions that are currently limiting the development of crops susceptible to frost, with an increase in the temperature level, will start to exhibit favourable conditions for the development of some crops. A typical case would be the coffee crop, which could in future be moved from the south-east to the south of Brazil (Pinto *et al.*, 2002).

Another aspect to be analysed is the direct effect on plants of an increase in the concentration of carbon dioxide in the atmosphere, which has been intensively studied by experts in plant physiology. The effect of photosynthetic activity and carbon dioxide concentration on plant growth is well known. The concentration of CO_2 in the atmosphere, at around 300 ppm (parts per million), is well below saturation for most plants. Excessive levels, close to 1,000 ppm, cause phytotoxicity. In this range, an increase in CO_2 generally promotes greater biological productivity in plants (Pinto *et al.,* 2002).

The consequent rise in sea levels, both due to the melting of glaciers and the expansion of the water mass in the face of the new temperature, will lead to the displacement of people located in coastal cities, approximately 40 per cent of the world's population, to the interior of the states.

The sum of global warming and habitat fragmentation could result in the extinction of many species of mammals and birds. Rising sea levels could jeopardise ecological security, including mangroves and coral reefs, natural nurseries for countless species.

Half of the mountain glaciers and large frozen areas could disappear by the end of the 21st century (Epstein, 2002).

1.3 WHAT IS GLOBAL WARMING?

Global warming is a climatic phenomenon caused by an increase in the Earth's average surface temperature. This is due to the high concentration of anthropogenic pollutants in the atmosphere, gases that intensify a natural phenomenon called the greenhouse effect, which has the function of balancing the planet's temperature. In this process, the Earth receives radiation emitted by the Sun and returns much of it to space in the form of radiation. GHGs (greenhouse gases), which are: water vapour suspended in the atmosphere, carbon dioxide, ozone, methane and CFCs, absorb infrared radiation emitted by the Earth's surface and radiate the absorbed energy back, so the surface receives twice as much energy as it would naturally receive from the Sun. These gases form a layer in the atmosphere that gets thicker and thicker, preventing the dispersion of energy from the Sun and also retaining all the infrared radiation emitted from the Earth's surface, thus increasing the temperature of the entire planet. Therefore, with the intensification of these gases comes global warming.

Large quantities of these gases have been emitted into the atmosphere since the Industrial Revolution began. Since 1750, carbon dioxide emissions have risen by 31 per cent, methane by 151 per cent, nitrogen oxide by 1 per cent and tropospheric ozone by 30 per cent.

It is believed that some of the emissions of these gases are due to natural causes such as volcanic activity, spontaneous forest fires and the decomposition of organic matter, but it is human actions that are most responsible for emissions due to cars, factory chimneys, fuel burning and other factors. After all, the extent to which human action can affect the planet's natural cycle is unclear.

The issue of the greenhouse effect and its influence on global warming, which has accelerated over the last 100 years, pits powerful social forces against each other that do not allow the issue to be dealt with from a strictly scientific point of view. On the one hand, there are the defenders of anthropogenic causes as the main culprits in the accelerated warming of the planet. They are the majority and omnipresent in the media.

> *"On the other side are the 'sceptics', who claim that accelerated warming is much more related to causes external to the Earth than to those claimed, deforestation and pollution causing the undesirable effects on life on the face of the Earth more quickly than the planet's replacement capacity" (Warming, 2006).*

However, global warming has generated a series of consequences, such as: rising sea levels, melting glaciers, ice caps and polar ice caps, changes in rainfall and wind patterns, intensification of the desertification process and loss of cultivated areas, as well as making phenomena such as hurricanes, typhoons, cyclones, tropical storms and El Nino (a natural phenomenon in the Pacific Ocean characterised by an abrupt change in the temperature of marine currents, causing major climate changes) more intense.

However, studies carried out with three kilometre-long cylinders have been set up to take ice samples from Greenland, which preserve climate records from the last 110,000 years, from which it is possible to estimate the temperatures that have already formed, arriving at results of sudden variations in temperature of up to 6°C. With these samples, they can imagine what might have happened around the world in those years. They also saw that during those times, various phenomena occurred all over the place: changes in temperature, strong winds, changes in rainfall, changes in heating and cooling, droughts and other factors, but they all happened in the same way, originating in a gradual change in temperature. These variations occurred several times.

> *Alley (2004) states that "most public policy discussions and research efforts in the area of climate change focus on global warming. But there's another problem: the climate has already changed suddenly in the past and will certainly do so in the*

Global warming should concern us more than ever: it could be facilitating the emergence of sudden changes in the Earth's climate, after all we can see climate change taking place all over the globe and especially in the oceans, which influence climate change on land.

1.4 OCEAN CIRCULATION AND ATMOSPHERIC INTERACTION

The ocean is an immense reservoir of heat, as it keeps the heat absorbed from solar radiation longer than the ground. When ocean water enters its great circulation scheme, heat is transferred vertically from the surface water to the ocean floor, back to the surface, and horizontally from a high latitude to a low latitude, from longitude to longitude. Thus, when the ocean releases heat in a region far from the point of absorption, this heat interacts with the atmosphere, moderating the daily and seasonal cycles and the temperature in areas of the ground surface (Simon & DeFries, 1992).

The convection scheme of water, and consequently the heat absorbed by it, has a major influence on energy dissipation and plays an important role in global climate dynamics (Simon & DeFries, 1992).

One of the properties of water is its high specific heat, which promotes a milder climate in places where the effect of the sea is greater. As the Earth's surface is almost entirely made up of water, it is of significant importance for maintaining the climate and life on this planet (Demillo, 1998).

The oceans are recognised as the main regulator of the climate. Ocean currents occur basically due to differences in water temperature and density. The oceans absorb heat and gases (e.g. carbon dioxide) from the atmosphere (Quadro *et al.*, 2004).

Vertical circulation, i.e. the movement of water at different depths, can bring "heat" and gases to deeper layers, which then remain "deposited" there for long periods of time. Ocean circulation patterns vary gradually over time, which can explain past climate fluctuations and variability (Gralla, 1998).

The oceans play an important role in the climate because they carry large amounts of heat from the tropics to the poles. They also store large amounts of heat, carbonates and CO_2 and are a major source of water for the atmosphere (through evaporation). The coupling of atmospheric and oceanic Global Circulation Models (GCMs) improves the physical realism of the models used to project future climate change, in particular the timing and regional distribution of changes (IPCC, 2001b).

Several models show only a marginal reduction in North Sea surface temperatures (in the North Atlantic) in response to the increase in greenhouse gases, relative to a slowdown in thermohaline circulation as the climate warms. This represents a negative local temperature *feedback.* The main influence of the oceans in climate change simulations is due to their large thermal capacity, which introduces a delay in warming that is not spatially uniform (Epstein, 2002).

1.5 WHAT COULD WARMING DO TO THE OCEANS?

As we have already seen, global warming can have various consequences for the planet, but its influence is mainly on the oceans, where their inhabitants have been affected by climate change.

As far as we know, rising temperatures affect the ocean currents that influence climate change. One of the most worrying variables is the rise in sea levels (Figure 1). Sea levels are rising by 0.01 to 0.02 metres per decade.

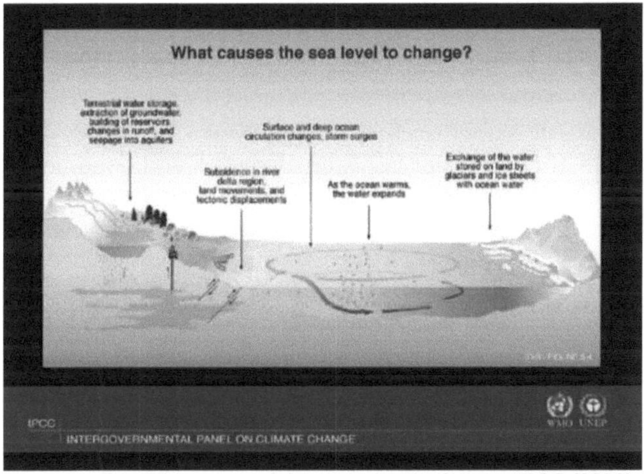

Figure 1 - Possible causes of rising sea levels. Source: IPCC (2006).

Figure 1 shows the causes of rising sea levels, including the shifting of tectonic plates, which can change the direction of marine and atmospheric currents, influencing changes to the climate, as well as the thermal expansion of ocean water and the melting of polar ice caps and glaciers, which is one of the biggest causes of rising sea levels.

> *Bindschadler and Bentley (2003) say that "ice can both accelerate and counteract the effects of global warming, because ice reflects much more solar energy back into space than dark oceans or land surfaces, causing part of the atmosphere above the ice to remain cold, which increases the possibility of more ice forming. On the other hand, the warming heats up the*

atmosphere enough to melt the ice, exposing more of the dark surface and thus absorbing more solar energy, making the air warmer".

Global warming could both slow down and speed up the rise in sea levels, because warmer air increases evaporation from the oceans, which contains more moisture than cold air. With warming, more seawater, evaporated from temperate zones, could be carried to polar zones and fall as snow. For this to happen, global warming would have to melt huge amounts of sea ice and expose more of the ocean's surface to the atmosphere. The problem is that warming could cause the ice to melt or break up more quickly.

Another important point concerns CO_2 emitted in large quantities due to the burning of fossil fuels, deforestation or fires, where part is absorbed by the oceans naturally, but studies show that this absorption rate has increased, causing chemical damage to the waters.

Changes are already taking place in the oceans, such as the Global Thermohaline Belt, which plays an important role in maintaining the climate in certain regions. The belt allows deeper waters to reach the surface, bringing nutrients and increasing the ocean's absorption of carbon dioxide. Worryingly, recent studies warn that there is already evidence of a slower circulation of the belt in the stretch of deep water between Scotland and Greenland.

"The dilution of ocean salinity - through the melting of Arctic ice, such as that of the Greenland ice sheet - and/or the increase in rainfall - could diminish or change the direction of the belt. This dramatic cooling could have dire consequences for agriculture and the climate in Europe, and impact ocean currents and temperatures around the globe" (Warming, 2006).

With regard to Greenland's glaciers, scientists on the Greenpeace ship Arctic Sunrise discovered in July 2005 that these glaciers were melting at high speed. Studies indicate that the Kangerdlugssuaq glacier, on the east coast of Greenland, is probably one of the fastest-moving glaciers in the world, with a speed of around 14 kilometres per year.

Furthermore, according to Greenpeace (2006), in 2005 the British Antarctic Survey revealed that 87 per cent of the glaciers on the Antarctic Peninsula have retreated in the last 50 years. In the last five years, the retreating glaciers have lost an average of 50 metres per year, and the entire Antarctic mantle has enough water to raise the planet's sea levels by approximately 62 metres.

In fact, rising sea levels can lead to: erosion of shorelines, coastal flooding and storm damage, saltwater contamination of drinking water supplies, increased salinity in estuaries, among other consequences for the oceans.

If warming continues in this way, it could lead to the extinction of several species and the faster the temperature rises, the more species will die out because they don't have enough time or can't adapt.

> *"The extinction of species is a real fact. For many species, the question is not whether extinction will occur, but when. In this context, anthropogenic activities have contributed to environmental degradation and altered atmospheric patterns and, due to the burning of fossil fuels, tonnes of carbon have been transferred from the biosphere to the atmosphere. Global warming, resulting from the process of emitting gases that contribute to intensifying the greenhouse effect, contributes to catalysing the process of extinction and consequently the loss of biodiversity"* (Simon & DeFries, 1992).

According to studies, thousands of marine animals, from the world's most exotic to the best known, could become extinct if Antarctic sea temperatures continue to rise as predicted, because the temperature of the waters in Antarctica has risen by one degree in the last 15 years. Studies have shown that species in areas with a rise of 1 to 2 degrees have already suffocated to death. However, the alarm goes out to the entire ocean, where species everywhere, especially the most sensitive ones, are in danger. Ecosystems such as coral reefs, which are extremely important for the oceans and contain a large number of species, suffer various consequences at high temperatures. This has the potential to have consequences for marine chains, not just at one point in the ocean, but for all marine fauna.

1.6 PREDICTIONS ON THE INFLUENCE OF GLOBAL WARMING

No-one has ever published a credible prediction of sudden climate change, nor is it expected to any time soon. This is because rapid changes are inherently more difficult to predict than global warming or other gradual changes (Alley, 2004). This is true, however, because we don't know for sure what may be causing or influencing all these changes in climate and how far they may go. What's more, as far as we know, several glaciations and sudden changes in temperature have already occurred and we're even close to one. We wouldn't be able to cause a new ice age, but the changes that could happen would be a terrible challenge for man and other living beings (Alley, 2004).

More and more studies are being carried out in order to get more answers. However, studies have predicted that certain places will experience changes in temperature. The Intragovernmental Panel on Climate Change (IPCC), a body linked to the UN, predicted in its latest report that average global temperatures will rise by 1.5°C to 4.5°C in the coming years (Alley; 2004).

Changes in climate currents could have consequences for temperatures around the globe. A global average rise in sea levels of between 9 and 88 centimetres is expected over the next 100 years, thanks to the greenhouse gases we have already emitted and will probably emit. Potentially, the West Antarctic ice sheet could contribute an additional six metres in sea level. Although the chances of such an increase happening are considered remote according to the Third Assessment Report of the Intergovernmental Panel on Climate Change (IPCC), recent research indicates new evidence of ice detachment from the ice sheet, and the entire Antarctic ice sheet has enough water to raise the planet's sea levels by 62 metres (Warming, 2006).

According to French scholar Paul Acot of the *Centre National de Recherches Scientifiques* (CNRS) in France, he has proposed two particularly worrying hypotheses: an increase in the average temperature of the globe of up to 4.0°C by 2050, affecting the northern hemisphere more because of the thermal inertia of the southern hemisphere, which is predominantly oceanic, and an increase of up to 1 metre or more in the level of the oceans over the same period. Knowing that 80 per cent of the world's population lives in coastal areas, the impact of this phenomenon can be assessed (Acot, 2003).

Professor Frédéric Bessat, from the University of Paris IV, in his article, the translation of which was recently published in Brazil (Bessat, 2003), based on climate models based on greenhouse gases and aerosols, identifies five trends for the next century, which would be as follows: a 2.0°C rise in temperature on average between 1990 and 2100; a rise in sea level of 0.50 to 0.80 metres by 2.100, with disruption to the ocean current circulation model; increased winter precipitation in the higher latitudes; intensification of the hydrological cycle (greater incidence of droughts and floods) and disturbances in the carbon cycle.

Entire species of marine animals and fish are directly at risk thanks to the rise in temperature, as they can't survive in warmer waters because they can't adapt.

In addition, with rising temperatures there is an increasing occurrence of diseases in marine animals.

Finally, all in all, we can see that we are not only facing studies that have yet to truly materialise, but facts that are occurring, global warming, changes in the oceans and their influence on species, and focusing on coral reefs because they are a great wealth of species, accounting for more than 65% of the entire ocean, and are being one of the ecosystems most affected by global warming.

CHAPTER 2

CORAL REEFS

2.1 THE CORALS

Corals are massive rock structures formed by the union of the skeletons of microorganisms and symbiotic algae; they are marine invertebrates (Figure 2).

The outside of the coral is made up of calcium carbonate, which binds the colonies to symbiotic algae (called zooxanthellae) that are embedded in the polyps and live inside them.

The polyps form tentacles with a mouth through which food is ingested with the help of tentacular urticating cells that release toxic substances into the water or directly onto the prey.

> *"Corals obtain nutritious substances produced by zooxanthellae, which also contribute to the deposition of calcium carbonate in their skeleton. The coloured pigments in the water also provide protection against ultraviolet rays. The algae obtain protection and use excretion products (carbon dioxide, ammonia, nitrates, phosphates) from their hosts to carry out photosynthesis and other metabolic processes."* *(Anemones, 2004).*

2.2 NUTRITION

Zooxanthellae are very important for the development and growth of corals, as they are responsible for their nutrition, the removal of metabolic by-products and the formation of their skeleton.

Corals feed on organic matter dissolved in the water, microorganisms and plankton. Those that feed on plankton are species with large polyps and no zooxanthellae.

2.3 REPRODUCTION

They reproduce asexually, which contributes to their expansion, or sexually, forming new colonies.

Even when the corals die, the polyps continue to serve as a base for new corals and any impact on the coral causes its death.

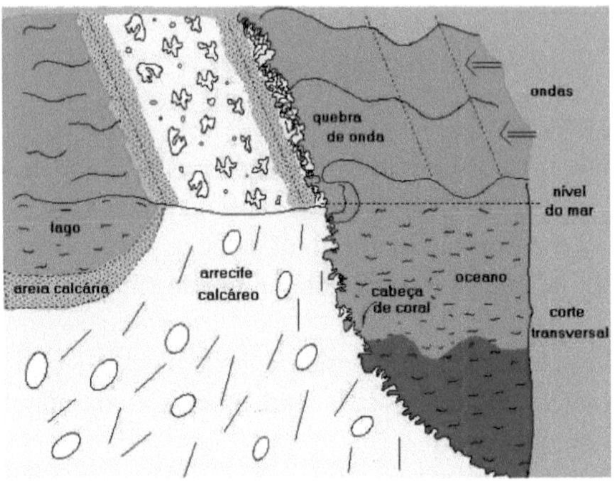

Figure 2 - Coral Reef. *Source: Ecology Engineering (2002)*

2.4 LOCATION

Its location depends on water temperature, light intensity, depth, suspended nutrients and current intensity (Figure 3).

The corals are located in clean, shallow waters no more than 50 metres deep, as the algae need light to carry out photosynthesis, and the water temperature can vary between 22°C and 26°C. The reefs grow vertically because they are distributed according to the intensity of the light.

Near the surface, where the waters are more agitated, there are branching corals like trees (fire coral is not a true coral but a hydrozoan, staghorn, elkhorn, among others). As the depth increases, the corals become globose, like the brain coral. Near the bottom, non-colony corals and gorgonians are typical.

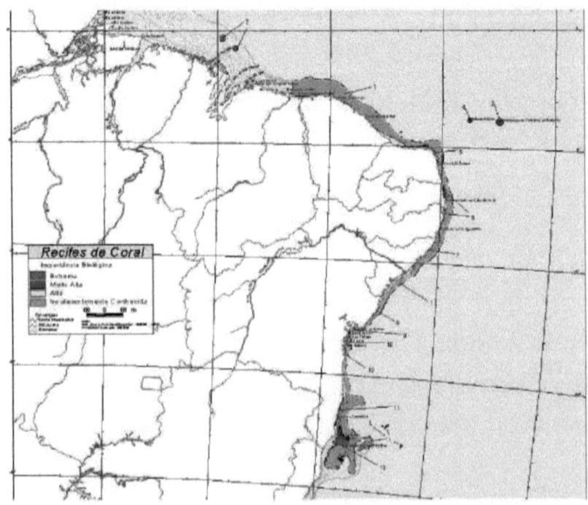

Figure 3 - Map of Coral Reefs in Brazil. *Source: Tropical Database (2005).*

Resistance to currents and the impact of waves varies according to the species. However, water circulation is important for their breathing, skeleton formation and feeding.

Reefs grow best in waters with few nutrients, as the water is clearer and allows better light penetration, and also does not favour the growth of macroalgae that could compete with corals for space and light. Some nutrients, such as phosphate, are harmful to corals because they prevent the formation of their skeleton.

Reefs protect the coast, as they break the force of sea waves, reducing marine erosion, they are responsible for producing organic matter and recycling nutrients, benefiting organisms that use them for shelter, reproduction and food, they are sensitive to temperature variations, growing vertically from 0.2 to 8 mm/year.

There are approximately 600,000 km^2 of corals in the world, of which 350,000 km^2 are located in Australia and form the Great Barrier Reef Marine Park, which is 2000 km long, made up of 2000 islands and 3000 coral colonies and also has a thickness of 1200 km between the volcanic base and the surface (Reefs, 2004).

In Brazil, 18 species of coral and four species of hydrocoral have been reported, which have their own characteristics, such as low species diversity, a lack of branching coral forms and species adapted to turbid waters.

Brazil has 4 to 8 gorgonian species, staghorn, green algae, anemone, daisy, sargassum, sea lily and brain.

In Brazil, coral reefs are distributed along some 3,000 kilometres of coastline, from Maranhão to the south of Bahia, and are the only reef formations in the South Atlantic Ocean.

Corals are home to around 30 per cent of the world's marine species and 65 per cent of fish, including moray eels, octopus, shrimp, starfish, crab, snails and sea slugs (Reefs, 2004).

In the south and southeast, the waters are cold due to the action of the Maldives (Falklands) current; while in the north and northeast, although the waters are warm, the mouths of large rivers such as the Amazon and the São Francisco cloud the water enough to prevent the development of large reefs.

Due to its irregular shape and the fact that it has many crevices and tunnels, various marine organisms use it as a home, which attracts larger species that go in search of food, and also causes the eggs and larvae of other organisms to fall between the crevices of the coral, protecting them from predators.

Thus, many fry seek out the reefs as a home, using them for protection and food, remaining on the corals until they reach adulthood, when they leave for the deep regions far from the coast. Corals are known according to their location: continental or oceanic. There are three types of coral and Brazil has a specific type.

Fringes are always close to shallow-water beaches and grow in a horizontal formation.

Barriers grow parallel to the coast and are always separated by a channel that can be shallow saltwater lakes. They form a buffer between the open sea and the beach.

Atolls can grow around sunken volcanoes. The islands reproduce a circular barrier structure, the best known being Bikini Atoll.

Mushroom-shaped Brazilian **hats.**

The largest coral reef is located in the Pacific Ocean and is called the Great Barrier Reef.

Open-brain corals (*Trachyphyllia* sp. Family Trachyliidae. Verril, 1901).

These corals live on the sandy bottoms surrounding coastal coral reefs. Here their ability to rise above the substrate by filling up with seawater prevents them from becoming covered by sand (Figure 4).

Figure 4 - *Trechyphyllia geoffroyi. Source: Pandora's Aquarium (2002).*

Bubble corals (*Plerogyra* sp. Family Caryophylliidae. Edwards & Haime,1848; Gray, 1847).

These corals have bubble-shaped vesicles that can reach considerable dimensions. These bubbles are not polyps, but tentacles for feeding. They develop in protected places with weak currents, where their vesicles can expand safely (Figure 5).

Figure 5 - *Pink Plerogyra sinuosa. Source: Home Aquária (2003).*

Carpet anemones (*Stichodactyla mertensii.* Family Stichodactylidae Brandt, 1835). This large anemone from the clear tropical waters of the Indo-Pacific lives on coral reefs. Its greenish colour is due to the microscopic algae (zooxanthellae) that live in its tissues. In this symbiosis, the algae benefit from protection, while the anemone obtains part of the nutritional substances they produce. The carpet anemone also offers protection to several species of clown fish (for example, *Amphirion ocellaris*), which are immune to the venom of its tentacles (Figure 6). It is the largest anemone in its family and can reach 1m in width (Anemones, 2004).

Figura 6   - *Heteractis magnifica* (Stichodactylidae). *Source: Water Worx (2003).*

CHAPTER 3
GLOBAL WARMING AND CORAL REEFS

3.1 THE CO_2 RATE AND BLEACHING

As CO_2 increases in the atmosphere, the coral skeleton becomes more fragile and the algae (zooxanthellae), which inhabit the corals and are responsible for their colour, become toxic when the water warms up, so they are expelled by the corals and coral bleaching occurs. The zooxanthellae die, but they are also important for the corals (they provide nutrients, oxygen and help to form the calcareous skeleton), so both will suffer (Camaguey, 2006).

According to Victoria Fabry, a biologist at California State University San Marcos and co-author of the journal Science (2004) *apud* OCEANO (2004), with the release and increase of CO_2 , there will be a decrease in the growth rate of limestone skeletons.

The problem is that the oceans absorb 48 per cent of the CO_2 emitted into the atmosphere, says Sabino, from the National Oceanic Atmospheric Administration (NOAA), the main author of an article on the subject published in the journal Science (2004) *apud* OCEANO (2004).

> *"According to Migotto from the University of São Paulo (2006), the high concentration of CO_2 , will lead to changes in pH and the saturation state of carbonates in the oceans, increasing the acidity of surface waters due to the higher concentration of carbonic acid, which could reduce the rates of calcium carbonate deposition by corals, affecting growth, but on the other hand, it should stimulate the growth and population increase of many algae, affecting the relationship between them and corals."*

Studies in 2004 revealed that the rate of shell calcification would drop by 25 to 45 per cent if the concentration of CO_2 reached 800 parts per million (ppm) in the atmosphere, whereas in 2004 it was close to 380 ppm. Increased acidity in the ocean can cause loss of shell calcification and change biodiversity, as organisms live better at certain pH levels.

Species will appear in certain regions and others will disappear, comments Ito, a specialist in dissolved gases in the oceans at the Oceanographic Institute of the University of São Paulo (USP).

The oceans are not infinite sources of CO_2 , they have absorption limits and according to the authors of the journal Science (2004) *apud* OCEANO, (2004), in hundreds of thousands of years, the oceans should absorb 90 per cent of anthropogenic carbon dioxide emissions (CO_2), but due to other mechanisms, in the short term, the ocean may lose its capacity to absorb CO_2 .

In any case, the high rates of CO_2 are having an impact on corals and other marine inhabitants, which will be at greater risk of extinction from now on.

3.2 HIGH TEMPERATURES AND CORAL REEFS

Corals are sensitive to changes in temperature, but when comparing when the temperature rises and when it falls, greater damage is seen when the temperature rises. This is how the phenomenon of bleaching occurs, which usually happens after a period in which the surface temperature of the sea water rises by a few degrees above the historical average for that particular period and location.

> *"Exposure to temperatures 4 to 5 °C above average for 1 or 2 days can be enough to cause gradual bleaching and lower mortality. Bleaching has occurred in areas warmed by the 'El Nino' phenomenon, but also in places and/or years not affected by it"* (Migotto, 2006).

Rising temperatures also have an impact on all marine food chains. Zooplankton, for example, which feed small crustaceans including krill, grow under the sea ice. A reduction in sea ice means a reduction in krill, which in turn feeds many whale species.

Whales and dolphins strand in high temperatures. Large whales could also lose their feeding area, the ocean around Antarctica, because of melting and collapsing ice sheets.

Entire species of marine animals and fish are directly at risk thanks to rising temperatures - they simply can't survive in warmer waters. Some penguin populations, for example, have declined by 33 per cent in parts of Antarctica because of habitat decline.

An increasing occurrence of diseases in marine animals is also linked to rising ocean temperatures (Warming, 2006).

Climate change is already taking place on our planet and the oceans are suffering. The greenhouse effect and holes in the ozone layer could affect coral reefs in several ways. One is the increase in temperature due to the greenhouse effect, which does not threaten the survival of reefs, but frequent episodes of extreme temperature peaks will increase the incidence of bleaching and mortality, making them vulnerable to other types of stress. The other cause is rising sea levels, associated with rising temperatures, which will not be beneficial for corals and those who depend on them, as there will be an increase in the rate of calcium carbonate deposition, when CO increases, which decreases the growth of calcareous plankton skeletons (Migotto, 2006).

3.3 HOW LONG WILL THE REEFS LAST?

If the rise in temperatures continues, coral reefs and the species that depend on them are at risk of disappearing. No one knows for sure what will happen to the reefs and those that depend on them. There is a need for more studies to identify and quantitatively understand these mechanisms so that it is possible to know what role the oceans will play as CO_2 storers in the future.

The average ocean temperature has already risen by around 0.5 °C in the last century. Warming threatens ten of the 18 largest coral reefs on the planet. Table 1 shows the level of risk of disappearance in the planet's main coral reef regions.

Table 1 - Level of disappearance of coral species.

Region	No. of species	Bass	Medium	High
Philippines	1488			X
Indonesia and Malaysia	1443			X
Japan	1262		X	
New Caledonia	1142	X		
Great Barrier Reef	1123	X		
India	1084		X	
Australia	824	X		
Mauritius and Réunion	762			X
Red Sea	679		X	
South Africa	572			X
Lord Howe Island	524	X		
Caribbean	450		X	
Hawaii	312	X		
California	145	X		
Gulf of Guinea	78			X
Cape Verde Islands	51		X	
Ascension Islands	40	X		
Easter Island	33	X		

Source: Revista época.

The following graphs show the condition of coral reefs in the main oceans and in South America (Figures 7 and 8).

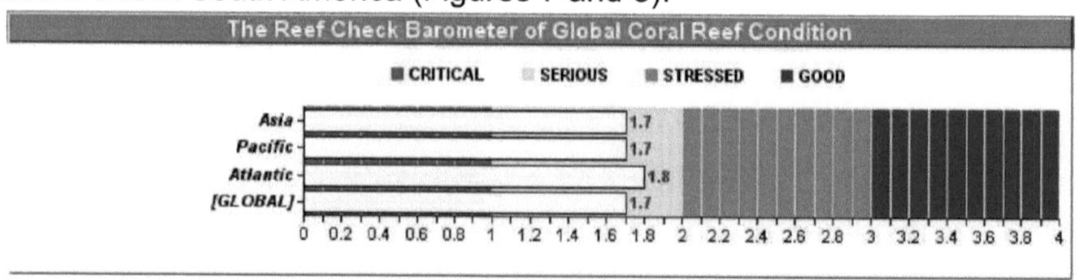

Figura 7 - The Reef Check barometer of the overall condition of the coral reef. *Source: data taken from Reef Check.*

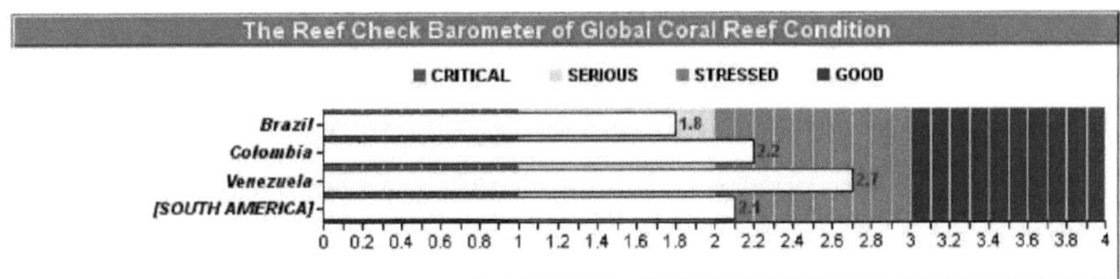

Figura 8   - The reef check barometer of Atlantic coral reef condition. *Source: data taken from Reef Check.*

CHAPTER 4
PRESERVING CORAL REEFS

4.1 IMPORTANCE OF CORAL REEFS

Considered one of nature's most extraordinary formations, coral reefs are an important marine ecosystem that is highly diverse on a local, regional and, above all, global scale. This is due to the fact that they are a habitat with an enormous diversity of plants and animals. They represent the largest centres of marine biodiversity, just like tropical forests for the terrestrial environment.

Coral reefs are rich in natural resources and of great ecological, economic and social importance, being responsible for the formation of beautiful natural pools, representing a tourist attraction worldwide (Figure 9).

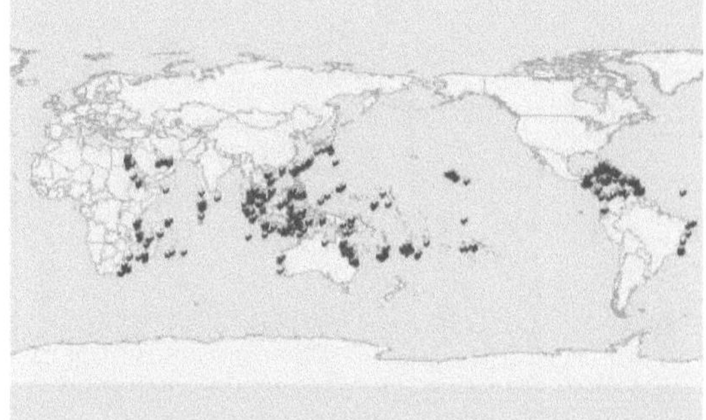

Figure 9 - Current map of the world's coral reefs. *Source: data taken from Reef Check.*

The fascination with corals is due to the fact that they are home to various species that depend directly on them for survival (at various stages of their lives, helping with marine equilibrium), that they form coral structures that protect the coast, that they form a natural dyke that prevents the sea from advancing, as well as being a source of food and income for many coastal communities, from fish of high commercial value to the variety and quantity of other species that are found on the reefs. To give you an idea of this variety, 105 species of fish have been found on coral reefs in Tamandaré alone. In addition, almost half a billion people live within 100 kilometres of a coral reef, and many depend on it for food and employment. In Brazil alone, 18 million people depend directly or indirectly on these environments. Around % of fish in developing countries comes from coral areas that provide food for approximately one billion people in Asia alone. Reefs protect beaches from erosion and help produce the fine sands that make them attractive for tourism, a main source of income for many tropical countries. In general, the goods and services generated by reefs were valued in 1997 at US$ 375 billion annually (Atlas, 2005).

Despite their great importance, Brazil still has little experience of organising the use of these ecosystems, especially given the alarming environmental degradation that coral reefs have suffered over the last century.

4.2 CAUSES OF REEF DEGRADATION

The degradation of reefs, which has been causing concern around the world, is closely linked to human and economic activities that have been taking place for some time.

One of the factors causing reef degradation is tourist activity. Studies carried out in these areas indicate that the main damage caused to corals located at shallow depths, just below the surface of the water, is generally caused by careless or inexperienced divers, who damage corals and other organisms by kicking them with their fins or even by standing on the reef.

For many reefs, it is likely that much of the damage observed can be attributed to uncontrolled tourist activities in the area. In a short period of time, members of the Tamar Project in Baía do Sueste observed hundreds of people walking on top of the reef, precisely over the habitat of the corals in question and various other invertebrates and algae.

After all, several examples given in various parts of the world indicate that tourism can only be compatible with the balance of an environment if it is carefully controlled.

Other influences on reefs are caused by fishing, an activity that is widely practised in coastal regions, especially in reef areas, as they contain a huge variety of species. This type of activity has been growing a lot, because in addition to its social importance, the product of fishing is the main source of income in coastal communities. Because of these factors, controlling overfishing has become a difficult task in order to prevent and preserve these sites. Other problems that threaten coral reefs are coastal water pollution and increased sedimentation rates caused by soil erosion.

In addition, the warming waves, probably the result of climate change, stress corals to the point where they expel the algae that inhabit them (zooxanthellae), leaving them "bleached". The bleaching of 1998, one of the hottest years in history, damaged huge areas of coral all over the world, seriously reducing the number of reefs. Nutrient and sediment pollution, sand and rock mining, the use of explosives and cyanide (or other toxic substances) in fishing, also stress the world's reefs.

4.3 REDUCING GLOBAL WARMING: THE KYOTO PROTOCOL

The Kyoto Protocol came about as a result of a series of events related to climate change, and is in fact an international treaty to reduce the emission of greenhouse gases, which if emitted in high concentrations cause global warming.

The protocol came into force on 16 February 2005, after several countries signed it. This treaty proposes that countries should reduce the amount of pollutants through actions that improve the observed situation.

The document establishes a reduction in emissions of carbon dioxide (CO_2), which accounts for 76% of all emissions related to global warming, and other greenhouse gases in industrialised countries.

If the Kyoto Protocol is successfully implemented, it is estimated that it should reduce global temperatures by around 0.02°C to 0.28°C by 2050. However, this will depend a great deal on the negotiations after 2008/2012, as there are scientific communities that categorically state that the 5.2 per cent reduction target in relation to 1990 levels is insufficient to mitigate global warming (Protocol, 2006).

Brazil signed the protocol's ratification letter on 23 July 2002 and is responsible for the annual production of 250 million tonnes of carbon
(10 times less than the US).

In 2001 in Bonn, Germany, at the protocol conference, the proposal of "carbon sinks" was suggested, where countries with large areas of forest that absorb CO_2 , could naturally use these forests as credit in exchange for controlling emissions.

However, it is necessary to carry out detailed studies on the amount of carbon that a forest is capable of absorbing, so that payments through carbon credits are not over- or under-valued. However, since the Johnesburg Conference, this proposal has become inconsistent with the objectives of the Treaty, which is to reduce the emission of gases that aggravate the greenhouse effect. Therefore, the policy should be to stop polluting, not to pollute where there are forests, as the balance would thus remain negative for the planet (Protocol, 2006).

Another project aimed at reducing global warming is the Clean Development Mechanism, which promotes development in underdeveloped countries. The aim is to encourage countries to generate clean energy, such as solar and biomass, to remove carbon from the atmosphere, where developed countries can invest in projects to reduce it and also use the credits to reduce their obligations, i.e. each amount that is not emitted or removed from the atmosphere can be purchased by the country that needs to meet targets for its reductions, giving rise to a global market for Certified Carbon Emission Reductions.

Another alternative, created by the US government in 2002, was the Clear Skies Initiative, the aim of which was to reduce the rates of the least significant pollutants causing global warming.

So instead of cutting the main greenhouse gas, carbon dioxide, the country would reduce emissions of three other gases by 2018, which together make up less than 15 per cent of the total components of the greenhouse effect. In the Clean Sky Initiative, economic growth comes first, and de-pollution follows (Entenda, 2005).

However, it can be seen that all the measures taken to reduce emissions under the Kyoto Protocol are suddenly linked to economic processes and not simply to environmental awareness or the future consequences of global warming.

4.4 TARGETS AND PRESERVATION AREAS IN CORAL REEFS

Recently, there has been a lot of discussion about the need to set up global monitoring programmes for coral reefs and other reef environments, due to the alarming degradation that has been observed in various parts of the world. According to the Intergovernmental Oceanographic Commission, this degradation occurs mainly in countries where specific environmental management programmes do not exist or have never existed.

Effective management programmes must therefore be created based on knowledge of the dynamics of reef ecosystems. To do this, it is necessary to obtain basic data on the biology and composition of reef species, data on the interactions of organisms, data on current levels and sources of impact or degradation and on the relationship between man and coral reefs.

However, until this information is obtained, in order to draw up an efficient project, the aim must be to reduce and limit any source of environmental impact as much as possible until the environment and possible sources of impact have been properly assessed.

Considering the importance of reefs, and with the concern about the degradation that has been occurring to Brazilian reefs, several projects and protection areas are being created.

CEPENE, the Oceanography Department of the Federal University of Pernambuco and the Peixe-Boi Centre have been working together for six years to research and protect corals, whose function in nature is comparable to that of the Amazon rainforest, due to the variety of animal and plant species whose life cycles are associated with them. This partnership culminated, in 1997, the International Year of the Reef, in the creation of the largest federal marine conservation unit in the country, the Coral Coast Environmental Protection Area (APA), the first federal conservation unit to protect part of the coastal reefs that are spread over some 3,000 kilometres of the northeast coast and the largest federal marine conservation unit in terms of extension, as shown in Figure 10 (Corais, 2003).

Figure 10 - Map of the Costa dos Corais Environmental Protection Area (APA). *Source: Ibama.*

The APA covers an area approximately 34 by 135 kilometres wide, covering 10 municipalities in the states of Pernambuco and Alagoas, from Tamandaré (PE) to Paripueira (AL). The initial partnership was expanded and now includes the environmental institutions of the two states and the 10 municipalities covered by the protection, as well as IBAMA's Manatee Project and the Marine Mammals Foundation (Corais, 2003).

Before the creation of the Coral Coast APA, only the reefs included in the areas of marine conservation units, such as Atol das Rocas, Fernando de Noronha and the Abrolhos Archipelago, were legally protected (Ecossistemas, 2006).

Since 1999, the Protected Areas Directorate (DAP) has been working specifically with this reef ecosystem, where various initiatives have been taken to establish a Coral Reef Protection Network.

The directorate began with a project, in partnership with the National Institute for Space Research (INPE) and the Coastal Reefs Project, to map the reefs inside Brazilian protected areas. The project: "STUDIES ON BRAZILIAN CORAL REEFS: TRAINING AND APPLICATION OF REMOTE SENSING MAPPING TECHNIQUES". This generated the "Atlas of Coral Reefs in Brazilian Conservation Units", containing maps of the entire Brazilian reef environment for the first time.

Throughout the area where the reefs occur, there are 9 marine conservation units, including federal, state and municipal ones, which cover the two groups of management categories: full protection and sustainable use. They are: Atol das Rocas Biological Reserve, Femando de Noronha National Marine Park, Abrolhos National Marine Park, Parcel do Manoel Luís State Park (MA) and Recife de Fora Municipal Park, Porto Seguro (BA). The sustainable use areas are: Costa dos Corais Environmental Protection Area (PE and AL), Ponta da Baleia State Environmental Protection Area (BA), Coral Reefs State Environmental Protection Area (RN) and the Corumbau Extractive Reserve (BA) (Ecossistema, 2006).

In addition, DAP launched the Campaign for Conscientious Conduct in Reef Environments (Figure 11), developed in partnership with the Coastal Reefs Project (BID/UFPE/IBAMA/FMM), with the support of the National Environmental Education Programme PNEA and IBAMA. (Atlas).

This campaign is part of the Campaign for Conscientious Behaviour in Natural Environments, which was promoted by the Ministry of the Environment (MMA) last year, with the aim of presenting rules, conduct and clarification to visitors in a protected area such as the reefs, thus aiming for fewer impacts in these areas.

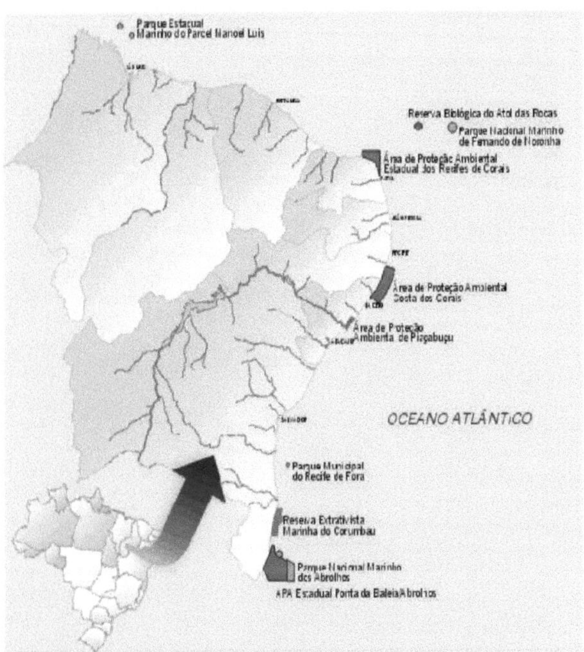

Figure 11 - Conservation units participating in the
Conscious Behaviour in Reef Environments Campaign. *Source: Ministry of the Environment.*

Other projects are emerging, such as the implementation of a sustainable plan for the reefs and the training of environmental agents and the organisation of basic courses in marine biology, monitoring and environmental education. The results are already beginning to emerge: the new Code of Ordinance for the city of Tamandaré, drawn up in 1997 with the participation of the two councillors-researchers from CEPENE/IBAMA, has Chapter II regulations on the marine environment, mangroves, beaches and their natural resources. It is very comprehensive and innovative. It has been in force since 26 January 1998 (Ecossistema, 2006). However, the creation of reef protection areas is not enough; it is also necessary, first and foremost, to raise awareness not only among the population, which has generally witnessed or enjoyed the reefs, but also among the entire world population about what is happening to coral reefs and the drastic predictions that could happen, in fact, expanding awareness projects and creating reef preservation areas in all parts of the world that contain reefs is one of the ways to maximise the recovery and preservation of this diverse marine ecosystem.

CHAPTER 5

CONCLUSION

Global warming and its consequences could cause irrecoverable damage to the oceans, and especially to coral reefs. However, it is believed that some information or questions still need to be clarified.

Little is still known about the marine environment, its organisms and its dynamics with the atmosphere, but it is clear that since the Industrial Revolution, man has been actively influencing the environment, accelerating degradation processes, favouring the extinction of species and changing the global climate. Further studies based on new technologies are essentially important in order to solve or at least minimise this problem.

It is also worth remembering that a massive effort to educate and raise environmental awareness is needed in all sectors of society. The population's contextualised awareness of environmental problems, the importance and current situation of reefs need to be on the agenda of all discussions on Public Environmental Policies, as well as, of course, being present in national curricular components.

Perhaps the creation of more environmental protection areas would be a start not only towards the recovery of degraded marine environments, but also towards the future preservation of one of the most biodiverse ecosystems on the planet.

BIBLIOGRAPHICAL REFERENCES

ACOT; (APUD) CONTI, J. B. **Considerações sobre as mudanças climáticas globais**. [s.l.: s.n].*Revista do Departamento de Geografia*, 2005.Available at: <http://www.geografia.fflch.usp.br/publicacoes/RDG/   RDG16/PDF/ Jos%C3%A9_Bueno_Conti.pdf>.Accessed on: 16 July 2007, 12:45.

ALLEY, R. B. **Abrupt Climate Change**. *Scientific American Brasil*, São Paulo, v. 3, n.31, p. 40-47, December, 2004.

anemones and corals. [s.l]: Oceanário de Lisboa, 2004. Available at: <http:// www.oceanario.pt/site/ol_dossier_03.asp?artigoid=119>. Accessed on: 02 June 2007, 19:13.

Global Warming. Greenpeace2006. [s.l.: s.n]. Available at: <http://oceans. greenpeace.org/en/our_oceans/climate_change>. Accessed on: 16 July 2007, 11:56.

Global warming. Wikipedia the free encyclopaedia, 2006. [Available at <http: //en.wikipedia.org/wiki/Greenhouse_effect>. Accessed on: 27 August 2007, 10:22.

ATLAS of Coral Reefs in Brazilian Conservation Units. [s.l]:Ministry of the Environment, 2005. Available at: <http://www.mma.gov. br/tomenota.cfm?tomenota=/port/sbf/dap/atlas.html&titulo=Atlas%20dos%20 Recifes%20de%20Coral%20nas%20Unidades%20de%20Conservação%20B rasileiras>. Accessed on: 20 September 2007, 10:50.

BECKER, F.D. , ALEMIDA, J. , GÓMEZ, W. H. , MULLER, G. , PHILOMENA, A.L. , RAMPAZZO, S. E. , REIGOTA, M. , VARGAS, P. R. , **Sustainable Development, Necessity and/or Possibility?,** UNISC, 3ª edition, Santa Cruz do Sul, 2001.

BESSAT; (APUD) CONTI, J. B. **Considerations on global climate change**. [s.l.: s.n]. *Journal of the Department of Geography*, 2005. Available at: <http://www.geografia.fflch.usp.br/publicacoes/RDG/RDG16/ PDF/Jos%C3%A9_Bueno_Conti.pdf>.Accessed on: 16 July 2007, 12:45.

BINDSCHADLER, R. A; BENTLEY, C. R. **Ghosts of the Thaw**. *Scientific American Brasil*, São Paulo, v. 1, n.08, p. 30-37, January, 2004.

CAMAGÚÉY, S. **Global Warming and the Impacts on the Coral Reefs**. Presencia Latina, 2006. Available at: <http://www.prela. nexus.ao/pag/noticia(6).htm>. Accessed on: 12 August 2007, 16:57.

CENAMO, M.C. , **Climate Change,** the Kyoto Protocol and the Carbon Market, 2004. Available at: www.cepea.esalq.usp.br/pdf/protocolo_ kyoto.pdf. Accessed on: 10 November 2004.

CONFEDERAÇÃO NACIONAL DA INDUSTRIA & CONSELHO TEMÁTICO PERMANENTE DE MEIO AMBIENTE, **Convention on Climate Change:** Considerations and Recommendations of the National Confederation of Industry. Available at: wwwsr.unijui.tche.br/ambienteinteiro/clima.pdf Accessed on: 01 December 2004.

CORAIS.[s.l]:Ibama,2003. Available at: <http://www.ibama.gov.br/projetos _centres/centres/cepene/corais.htm>. Accessed on: 30 September 2007, 20:35.

DEMILLO, R. , **How Climate Works,** Quark, São Paulo, 1998.

DIAS, M.S.F.S., **Interação Biosfera Atmosfera Em Mesoescala Na Amazônia,** Departamento De Ciências Atmosféricas Instituto De Astronomia, Geofísica E Ciências Atmosféricas Universidade De São Paulo, Relatório à FAPESP, 2001. Available at :http://lba.inpa.gov.br/ Accessed on 1st December 2004.

DIVERSITY at risk. São Paulo: Época magazine, 2006. Available at: <http://revistaepoca.globo.com/Epoca/0,6993,EPT731691-1655-5,00.html>. Accessed on: 21 July 2007, 16:53.

Coral Reef ECOSYSTEMS. [s.l]:Viva Brasil,2006. Available at:<http://www.vivabrazil.com/recifes_coralineos.htm>. Accessed on: 20 September 2007, 15:20.

ENGINEERING Ecology. [s.l]: State University of Campinas,2002. Available at: < http://www.unicamp.br/fea/ortega/eco/iuri10b.htm >. Accessed on: 01 October 2007, 15:35.

UNDERSTANDING the Kyoto Protocol. [s.l]:Época,2005. Available at: <http://revistaepoca.globo.com/Epoca/0,6993,EPT908417-1655-1,00.html>. Accessed on: 30 October 2007, 20:25.

EPSTEIN, I. **Global Warming**, Com Ciência Revista Eletrônica de Jornalismo Científico, 2002. Available at: http://www.comciencia.br/ reportagens/framereport.htm Accessed on: 01 December 2004.

GRALLA, P. , **How the Environment Works**, Quark, São Paulo, 1998.
HETERACTIS magnifica (Stichodactylidae) - Anemone. [s.l]: Water Worx,2003. Available at: <http://www.waterworxbali.com/anemone.shtml>. Accessed on: 02 November 2007, 21:35.

IPCC. **Intergovernmental Panel on Climate Change**. 2001. Climate Change 2001: Impacts, Adaptation and Vulnerability. Working Group II. TAR: Summary for Policymakers. Available at: www.meto.gov.uk/sec5/ CR_div/ ipcc/wg1/WG1-SPM.pdf Accessed on: 25 November 2004.

IPCC. **Intergovernmental Panel on Climate Change**. 2001. Working Group I. Third Assessment Report. Summary for Policymakers. WMO. 17 pp. Available at: www.meto.gov.uk/sec5/CR_div/ipcc/wg1/WG1-SPM.pdf

Accessed on: 25 November 2004.

LORA, E.E.S. , **Prevention and Control of Pollution in the Energy, Industrial and Transport Sectors**, Interciência, 2ª edition, Rio de Janeiro, 2002.

MAPA Apa. [s.l]:Ibama,2003. Available at: <http://www.ibama.gov.br/ projetos_centros/centros/cepene/mapao.htm>. Accessed on: 30 October 2007, 21:05.

MAP Recife. [s.l]: Tropical Database,2005. Available at: <www. bdt.fat.org.br/.../ mapas/recifes_a3.jpg>. Accessed on: 03 June 2007, 15:30.

MARC, P.S. , **A Poluição,** Salvat, Rio de Janeiro, 1979 Mazza, P., Roth, R., **Global Warming Is Here: The Scientific Evidence**, Atmosphere Alliance/Energy Outreach Centre, 1999. Available at: www.newconnexion. net/article/07-99/global.html Accessed on: 01 December 2004.

MENDES, J. C. , **Basic Palaeontology,** University of São Paulo, São Paulo, 1988.

MIGOTTO, Alvaro Esteves. Coral reefs and "bleaching". São Paulo: USP, 2006. Available at: <http://www.usp.br/cbm/artigos/ branqueamento.html>. Accessed on: 12 August 2007, 19:02

MULLER, P. B. **Bio Climatologia**, 2nd edition, Sulina, Porto Alegre, 1982.

OCEAN absorbs 48 per cent of CO_2 **released into the air**. Jornal do meio ambiente, Rio de Janeiro, July 2004. Science. P. 17.

ODUM, E. P. **Ecologia**, Guanabara, Rio de Janeiro, 1988.

PEREIRA, A.S. , **Mudanças Climáticas e Energias Renováveis**, Com Ciência Revista Eletrônica de Jornalismo Científico, 2002. Available at: www.comciencia.br/reportagens/framereport.htm Accessed on: 01 December 2004.

PINTO, H. S. , Assad, E. D. ,Zullo, J. Jr, Bruini, O. , **Global Warming and Agriculture**, Com Ciência Revista Eletrônica de Jorna lismo Científico, 2002. Available at: www.comciencia.br/reportagens/framereport.htm Accessed on: 01 December 2004.

PLEROGYRA sinuosa rosa. [s.l]: Home Aquaria,2003. Available at: <http://www.homeaquaria.com/public/catalog/index.php?cPath=242>. Accessed: 02 June 2007, 21:15.

QUADRO, M. F. L. , Machado, L. H. R., Calbete, S. , Batista, N. N. M. **Climatology, Precipitation and Temperature,** Centre for Weather Forecasting and Climate Studies - CPTEC/INPE. Available: www.cptec.inpe.br/products/ climanalise/cliesp10a/chuesp.html Accessed on: 01 December 2004.

RECIFES de Corais. Pernambuco: Museu do Una, 2004. Available at: <http://www.museudouna.com.br/corais.htm>. Accessed on: 23 October 2007, 20:08.

ROSA, L.P. ROVERE, E. L. ,**Overview of Latin American Technology Development for Avoiding Greenhouse Gases Emissions and for Mitigating Climate** Change, Global Change Centre, 1998. Available at: www.ivig.coppe.ufrj. br/doc/baires.pdf Accessed on: 01 December 2004.

SIMON, C. DEFRIES, R.S. , **One Earth, One Future,** MAKRON Books, São Paulo, 1992.

TEIXEIRA, W. , TOLEDO, M.C.M. , FAIRCHILD, T.R. TAIOLI, FABIO, **Deciphering the Earth**, Oficina de Textos, São Paulo, 2000.

The reef check barometer of global coral reef condition California: Reef check, 2006. Available at: <http://www.reefcheck.org/datamanagueta>. Accessed on: 14 September 2007, 19:32.

TRECHYPHYLLIA geoffroyi. [s.l]: Pandora's Aquarium, 2002. Available at: <http://www.fishpalace.org/T_geoffroyi.html>. Accessed on: 02 June 2007, 20:50.

What causes the sea level to change? IPCC Intergovernmental Panel on Climate Change. [s.l.: s.n]. Available at <http://www.ipcc.ch/present/graphics.htm>. Accessed on 16 September 2007, 11:15.

WILSON, E.O. FRENCH, M.P. **Biodiversidade**, Nova Fronteira, 2ª edition, Rio de Janeiro, 1997.